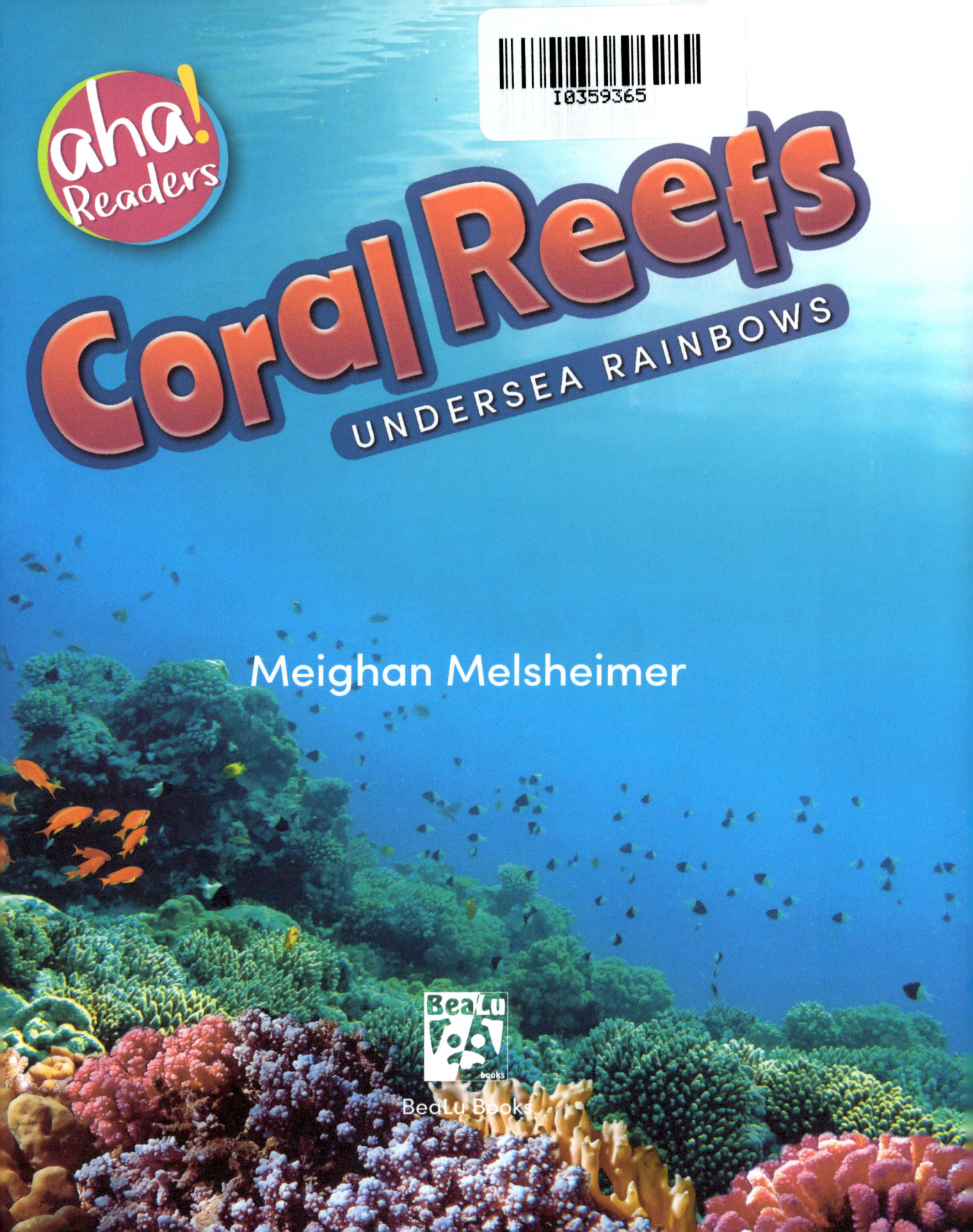

Copyright © 2020 by Meighan Melsheimer

All rights reserved. No part of this publication may be reproduced in any form or by any electronic or mechanical means, including information storage and retrieval systems, without the express written permission from the publisher, except in the case of brief quotations embodied in critical articles or reviews. For information regarding permission, contact BeaLu Books.

ISBN Hardcover: 978-1-7341065-2-7
ISBN Paperback: 978-1-7333092-5-7

Library of Congress Control Number: 2019952464
Publisher's Cataloging-in-Publication Data is on file with the publisher.

Edited by: Luana K. Mitten
Book cover and interior design by Tara Raymo • creativelytara.com

Printed in the United States of America
October 2019

BeaLu Books
Tampa, Florida

www.BeaLuBooks.com

PHOTO CREDITS: Cover: © John_Walker; Page 1: © Vlad61; Page 3: © Drew McArthur, © Siberian Art; Page 4: © Rostislav Ageev; Page 5: © Damsea; Page 6: © Designua; Page 7: © scubaluna; Page 8: © tank zoobar; Page 9: © unterwegs; Page 10: © Irina Markova; Page 11: © scubaluna, © Choksawatdikorn; Page 12: © Chainarong Phrammanec; Page 13: © Richard Whitcombe, © Ethan Daniels; Page 15: © Ethan Daniels; Page 16: © Richard Whitcombe; Page 17: © Rido, © Sergey Novikov, © Monkey Business Images

Table of Contents

What are coral reefs?.................... 4

How are coral reefs formed?............... 6

What do coral reefs need to survive?......... 8

Why are coral reefs so colorful?............12

What would happen if coral reefs
　　disappeared?..........................14

What are coral reefs?

Coral reefs are amazing **habitats** teeming with life! A rainbow of colors can be found in coral reefs where **exotic** plants and fish live and grow.

Did you know that more than one million **species** of animals and plants live on or near coral reefs? Look closely. What do you see?

Did you know that chemicals in some sunscreens can lead to bleaching in corals? When this happens the bright colors that you see in this picture are gone and the coral could die.

How are coral reefs formed?

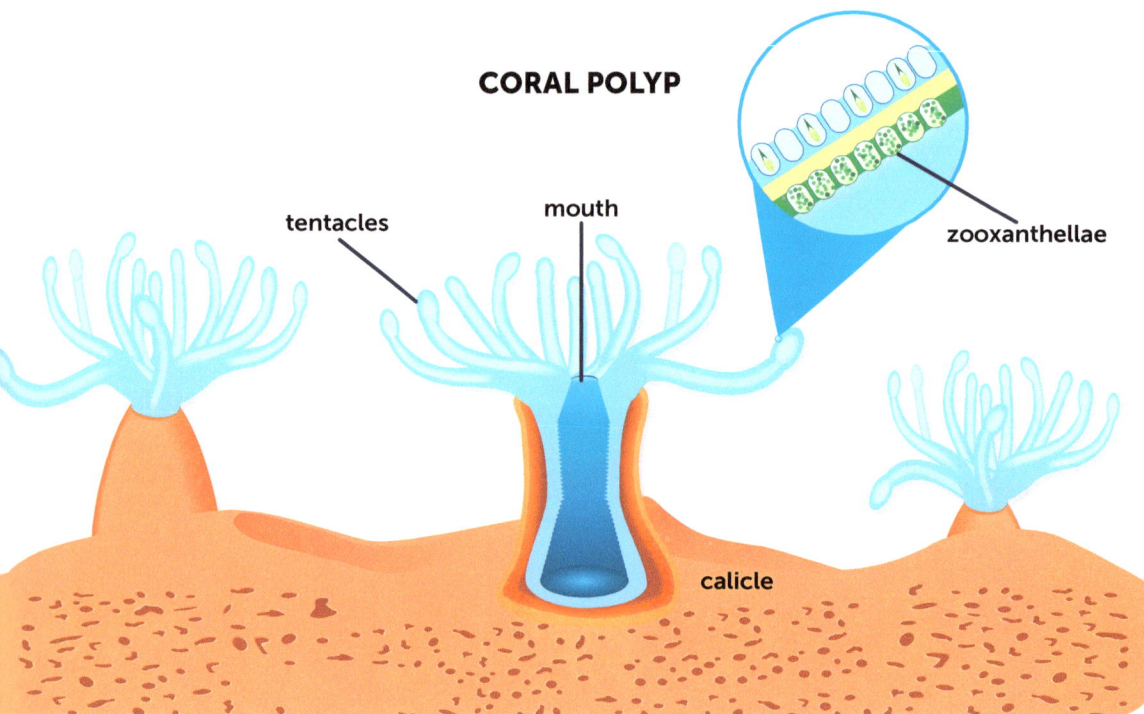

Coral **polyps** are teeny-tiny organisms that have a soft body and a hard **limestone** base. This base, called a **calicle**, makes the hard coral reef structure.

A coral polyp must first attach itself to a solid base. When the coral polyps group together, they create a **colony**. As colonies grow and develop for hundreds, even thousands of years, they form a reef.

Coral reefs grow extremely slow. They may grow anywhere from half an inch to seven inches in a year.

What do coral reefs need to survive?

Warm, salty, and clean, that's how coral reefs like their water. This helps the **zooxanthellae** that live inside of them to survive.

Zooxanthellae are algae that grow inside of each coral polyp. This algae needs sunlight to remain alive.

Reefs thrive in waters with waves because waves bring food into the habitat.

Coral polyps come out of their skeleton to feed, stretching their long bodies to capture zooplankton.

zooplankton

In addition to the zooxanthellae that live in their bodies, coral eats by catching tiny floating animals called **zooplankton**.

Why are coral reefs so colorful?

Many animals depend on coral reefs. Some experts estimate that over one million plant and animal species are connected to reefs. No wonder a rainbow of colors can be found on coral reefs!

Coral reefs provide shelter for many animals. The most common types of animals that live on coral reefs include jellyfish, eels, sea anemones, and fish, like the clownfish.

moray eel

jellyfish

sea anemones and clownfish

What would happen if coral reefs disappeared?

Safety and protection are provided by coral reefs. Scientists **hypothesize** that if coral reefs disappear the animals that live there would have to try to find new homes. Many of these animals would die.

This could directly impact us too. People who work in the fishing industry would no longer have jobs. Countries that depend on families to spend their vacations snorkeling and diving around coral reefs would lose money. People's health would suffer because doctors would no longer be able to use coral reefs to discover **remedies** to help people that are sick.

Coral reefs are dying because of warmer water temperatures caused by global warming and pollution.

Coral reefs are incredible! They provide a home for millions of diverse and different marine animals. They help us by providing beautiful places to visit, job opportunities, and remedies for illness. Without coral reefs in our **ecosystem**, life would not be the same. It is up to us to protect our coral reefs—our undersea rainbows!

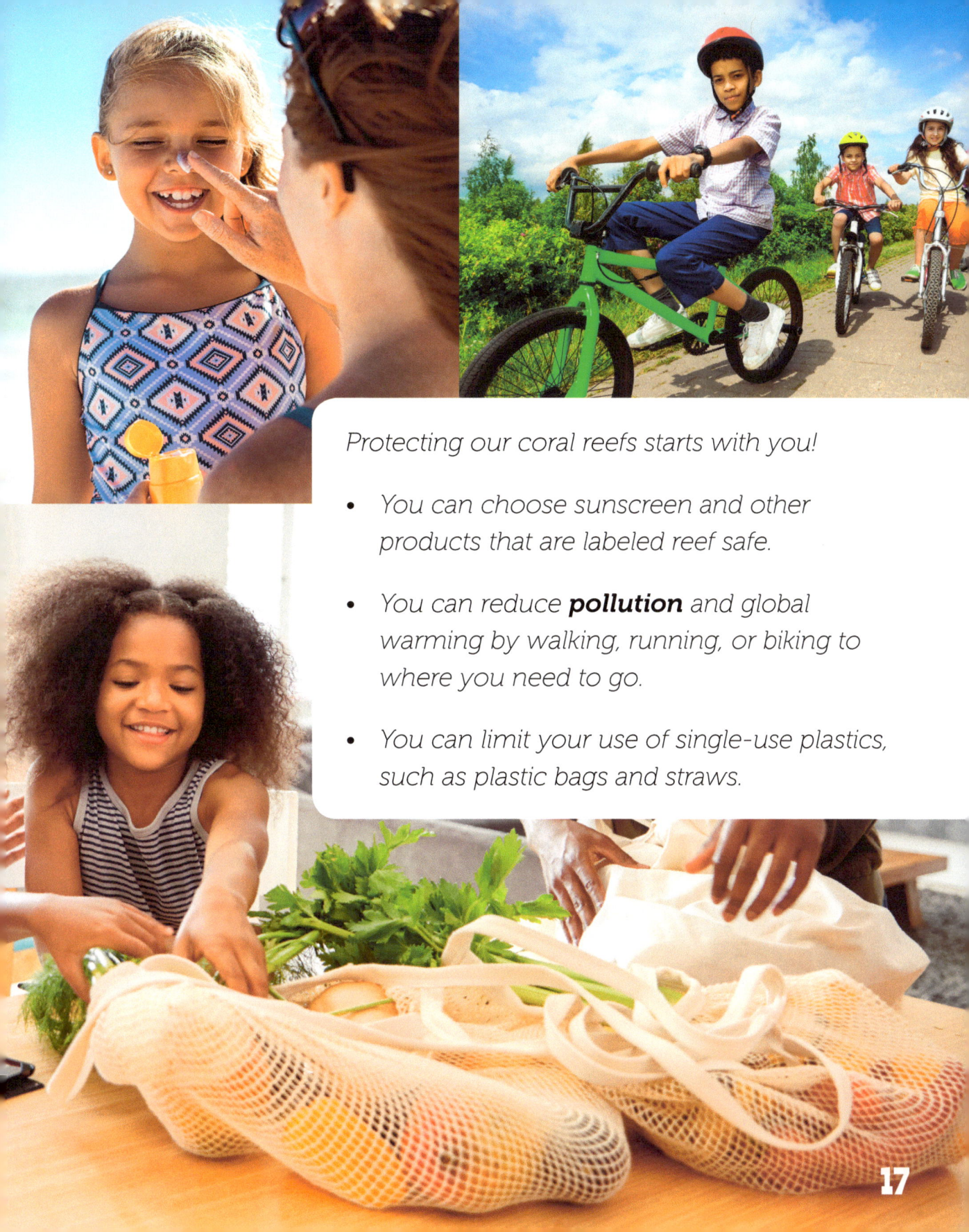

Protecting our coral reefs starts with you!

- You can choose sunscreen and other products that are labeled reef safe.

- You can reduce **pollution** and global warming by walking, running, or biking to where you need to go.

- You can limit your use of single-use plastics, such as plastic bags and straws.

About the Author

Meighan Melsheimer has spent her career as an educator. She lives in Florida and is an advocate for the environment. She enjoys spending her free time visiting Florida's beautiful, sunny beaches, and snuggling with her puppies.

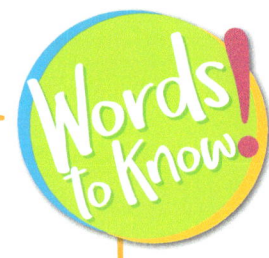

calicle (KAL-i-kul): a coral formation

colony (KAH-luh-nee): many coral polyps grouped together

conserve (Kuhn-SURV): to prevent the harmful overuse of something

ecosystem (EE-koh-sis-tuhm): a community of living organisms

exotic (ig-ZAH-tik): something that unusual or is striking

habitats (HAB-i-tatz): the natural home or environment of an animal or a plant

hypothesize (hi-POTH-uh-sahyz): to make an educated guess

limestone (LIME-stohn): hard, sedimentary rock that is often used as building material

pollution (puh-LOO-shuhn): a substance that is harmful or poisonous to an environment

polyps (PAH-luhps) a small growths, typically with a stalk or base

remedies (REM-i-deez) treatment for a disease or injury

species (SPEE-sheez) a group of similar organisms able to exchange genes

zooplankton (zoh-uh-PLANGK-tuhn): small, microscopic animals

zooxanthellae (zoh-uh-zan-THEL-uh): microscopic, single-celled organisms

www.ingramcontent.com/pod-product-compliance
Lightning Source LLC
Chambersburg PA
CBHW041125070526
44584CB00003B/281